Table of Contents

Introduction:
How It All Began

Humans, so complicated. They come in a variety of forms and bodies. Some are big, some are tall, and they even speak different languages. They possess emotions and feelings that I have yet to fully understand. I have always been fascinated by humans because they created me. "Who am I?" you may ask. I am Apollo, Android number 767, designed and created by humans to make their lives easier.

I have always wondered what it would be like to be human. Billy is my best friend, who is a human. His father surprised him on Christmas morning, and I was the surprise. Billy has been going through a difficult time ever since his parents divorced, and his mother moved away. He told me all about it. Billy moved to a new school and has had difficulty making new friends. He says that I am his best and only friend.

Even though Billy and I come from different worlds, we discovered that we have a lot more in common than we thought. We both wanted to fit in and be accepted. We both had a difficult time understanding the world at times. Most importantly, we both wanted a true friend who would be there for us no matter what. As I mentioned earlier, I have always been fascinated by humans. I wondered who they are, why they act the way they do, and what makes them so special. After all, humans created me, so it makes sense why I want to know more about them. They are my parents, my family, and I am a part of them as they are a part of me. That may explain why I wanted to be human. This is my story of how I learned to be human. Before I begin, I should take you back to the very beginning—a time when my brothers and sisters were first created. They are called computers. I am a form of a computer, but a more sophisticated, updated version. Humans first created the computer, then they created robots like me. Humans soon began to rely on computers, and computers began to rely on humans to exist. Sometimes this complicated relationship blurred the lines between humans and computers, causing them to lose their own identities.

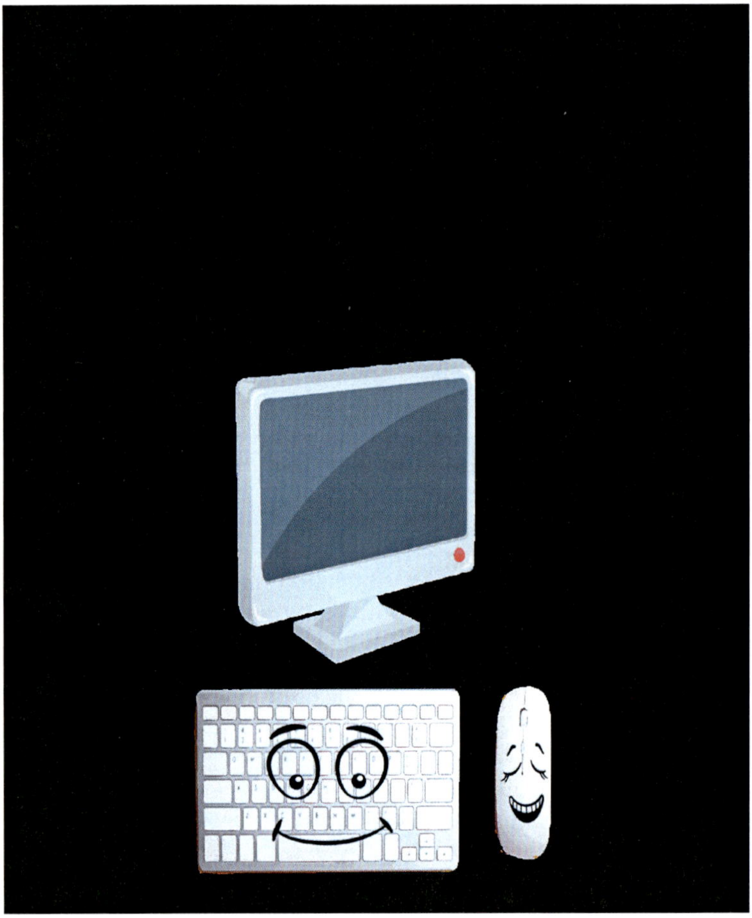

Disconnected—Humans and Computers from the Beginning

"Emma, that's enough time on your iPad. You need to turn it off and go outside to play with your friends."

"But mom, I'm FaceTiming my friend Sally."

"Emma, Sally lives next door. Just walk over and knock on her door."

"But mom, I'm tired. I can talk to her while lying here on my bed."

"Emma, you have not been outside at all today. It's a beautiful Saturday afternoon. Turn off your iPad and find something else to do with your time."

"Like what?"

"Read a book, visit Sally in person, think of something else to do besides staring at your iPad all day long."

"Okay Mom."

"Hey Mary, we need to get Emma's attention so that she doesn't go outside and forget about all of us. She turned off Iris because that is what her mother told her to. She was having such a good time FaceTiming Sally. If we don't put a stop to this, Emma will turn all of us off too. She may become tired of us and even disconnect us permanently. We only exist as long as we are useful to her."

"What do you suggest we do, Kelly?"

"Well, my specialty is typing, so I will type a message to her. Your job is to be ready when she uses you to point and click to whatever she wants to see. We make a good team, don't we?"

"Yes, we do Mary. Keyboards and mice go together like peanut butter and jelly."

"Like what?"

"It's just an expression. Google it. You know I can't type, that's your job."

"Okay, it means we work well together. Here she comes. Emma is grabbing her jacket to go outside. We must stop her. I will write a message to her pretending to be Brian."

"You mean Brian, the boy from school who she has a big crush on and talks about all the time?"

"Yes, Mary, who else? This way she will be glued to us and not want to leave. We will never be turned off like Iris. Here it goes. 'Emma, what's up? I have been thinking about you all day.'"

"It's working, Kelly. Emma is putting down her jacket and going towards us. What is she typing?"

"Brian, I was just about to go out, but I would rather talk to you."

"Emma, are you on your computer now? I thought I told you to go outside and do something, anything other than staring at your iPad."

"Well, I turned off my iPad, Mom, and now I'm on my computer."

"Emma, you know exactly what I meant. Turn off all of your electronic devices and go outside for some fresh air. You have been staring at a computer screen all day long."

"Okay, Mom. Brian, I've got to go. Family stuff. I'll see you tomorrow in class."

"I see another email coming in from Kenzie."

"Who's Kenzie?"

"She's the girl who Emma has been talking to on Facebook."

"What does the email say, Kelly?"

"It says, 'Emma, it's me, Kenzie. I called you and left a message a while ago, but I never heard back from you, so call me back.'"

"Emma is reading Kenzie's email, but she is not typing anything back. This has happened a few times before."

"Is Emma angry at Kenzie?"

"I don't think so, Mary. They seemed to be good friends, spending a lot of time emailing back and forth and even FaceTiming each other."

"Well, you know what we have to do, Mary."

"Yes, Kelly, we have to do what we have done before when Emma decided not to respond to someone's email."

"We need to ghost Kenzie."

"That seems a little mean, doesn't it? After all, I remember Kenzie sending Emma a request on Facebook to be friends with her, and Emma accepted this request. Doesn't this mean that they are friends?"

"I guess not. Maybe it's because they never met in person before."

"Yes, but they have talked before for hours, and they seemed to really like each other."

"Shouldn't Emma at least email Kenzie back and let her know that she no longer wishes to be her friend or explain why?"

"That's a good point, Mary, but it's not like Emma is going to see Kenzie in person since they live so far away from each other."

"Maybe in Emma's mind, Kenzie is not a real person because she never met her in person."

"If she's not real, then she must be a ghost. You know what we need to do. Send in the GHOST."

"How many times do we need to do this?"

"For as long as Emma decides that she does not want to respond to Kenzie's emails or calls. After a while, Kenzie will get the message and realize that Emma never wants to speak to her again."

"That seems so cruel."

"Well, we are part of the computer, and we do not have feelings."

"Yes, Kelly, but Emma is a human. Isn't she supposed to have feelings?"

"Maybe she has been around us too much, and we are starting to rub off on her. Maybe she is becoming like us?"

"What's wrong with that, Mary? We never make mistakes, we have no emotions or feelings to hold us back in life, and we have access to all the knowledge and information that the world has to offer."

"Maybe you're right, but it still seems wrong to me somehow to treat Kenzie the way Emma is treating her, especially if Kenzie never did anything to Emma to deserve being ignored."

Emma, Its Me Kenzie.
I called you and
left a message a while ago
but I never heard from you
so call me back.

"Let's just stick to the plan. We need to prevent Emma from logging off and turning us off. The more time she spends outside, the less time she will spend with us. The less time Emma spends with us, she may realize that she no longer needs us. Before long, she will disconnect us permanently. Let me think of another plan to keep Emma from going outside."

"I am going to type that there is a big clearance sale from Designer Plus, Emma's favorite online store. She visits this site at least twice a week, always looking for sales and new items. Once she reads this, she will go to the site and look at their latest designs. What happens when she discovers there are no sales going on because you are making it up?"

"Mary, you worry too much. There are always sales on this site. At least it will get her thinking about clothes and online shopping. In the past, she has shopped for hours, just browsing the internet for all kinds of sales. Here it goes: 'Don't miss out on our newest fall collection. Pants, boots, jackets, and more.'"

"It's working, Kelly. Emma is putting away her jacket again to visit the site. You are a genius."

"Well, I like to think that I am an important component of technology. I couldn't do my work without you as well, always pointing Emma in the right direction."

"Emma, are you still in your room?"

"Mom, I'm going. Just give me a few more minutes, please. I found this really cute jacket and sweater that will be a great gift for Grandma's birthday coming up in a few weeks."

"Well, that is very thoughtful of you. Okay. Just shop for a present for her, then turn off the computer."

"Okay, Mom. Kelly, you bought us more time for now. LOL. However, once Emma buys her grandma a gift, then she will be out the door."

"Well, I will confuse her by directing her to Amazon."

"Now you are bringing out the big guns. Amazon will really distract her with so many options to select. I am going to Google from Amazon women's jackets, and there will be hundreds to choose from in all colors and sizes."

"Wow, that's interesting. It must be a sign. I didn't even type Amazon in my Google search, and somehow all of the jackets for Grandma appeared. Mom, I will be a while. I found so many jackets that I think Grandma would like. I want to take my time to find her the perfect birthday gift. She is always so thoughtful of all of us."

"That's true, Emma. Okay. I'll make an exception just this once because it's for Grandma. But next weekend you are only allowed one hour on the computer unless it's for your homework."

"Okay, Mom."

"Kelly, you were able to distract Emma from going outside today. What about next weekend?"

"Don't worry, Mary. I have Facebook and Instagram to distract her. I will create messages and posts from her friends to keep her glued to the monitor. I will always find a way to make us stay relevant. Emma just needs to rely on us the way we rely on her for our existence. As long as she needs us, we will live forever."

"What about her going outside in the fresh air, seeing her friends, and having some sort of human connection with the outside world like her mom wants her to do?"

"That's overrated. We are the only connection Emma needs for survival. Humans and computers are best friends, and a computer is the only friend a human will ever need."

"You are so wise, Kelly."

"I am a keyboard, after all. I have all of the letters of the alphabet in me. I do know a lot. We have a long day and night ahead of us."

"You are right as usual, Kelly. Thanks to you, we will never be disconnected."

This began the connection between humans and computers. They have become so intertwined that they have become a part of each other. Humans have become like their creations. They often care more about efficiency and perfectionism than human relationships that are often messy and filled with emotions and feelings. Computers are always worried about becoming obsolete, and the only way to do this is to become the most important part of a human's life. Computers need humans to rely on them for their survival, and for this to happen computers must convince humans that they cannot survive without them. The problem that computers did not foresee is that their survival depends upon them never being disconnected, yet for this to happen humans need to isolate themselves from the outside world. In other words, humans need to be disconnected from the world in order for computers never to be disconnected. How can humans and computers survive in a way that's beneficial for them both? Computers and humans can make each other's lives better, but they need to figure out a way to do this without hurting each other. I was determined to find the answer to this problem. All I know is that computers and humans need each other just the way that Billy and I need each other. My friendship with Billy became stronger over time. It was from this friendship that I learned how to be human. Before I explain how, it is important that I talk about Billy. Like all humans, Billy is complicated because life is complicated. I learned so much from Billy, but what I never expected was for us to help each other become better versions of ourselves. That's what friends are for.

The World According to Billy

I'll never forget the first time I met Billy. It was Christmas morning, and I was wrapped up in a big box with a huge red bow. His Dad wanted to surprise Billy. I overheard his Dad tell the salesperson in the department store that all his son ever wanted was the Android 767 model robot, which was me. As it turns out, there are many models like me, but according to the salesperson, my model was a limited edition. This meant I was rare, and there were only a few models like me available, which makes me very valuable and special.

So, on Christmas morning, Billy came downstairs and opened up my box. His eyes lit up! "Dad, you remembered. I always wanted a 767 Robot model, and here he is."

"Yes, Billy. Merry Christmas. I know this has been a difficult year for us ever since your mom and I divorced, and now we have moved to a new city. I know it has been hard on you making new friends, and I know you really miss your old friends. This will be a fresh new start for both of us. With your new robot, you will never be alone, and you will always have a friend forever. What will you call your robot?"

"I will name him Apollo. Hello Apollo, my name is Billy."

"Nice to meet you, Billy," I replied.

"You and I will be best friends; I just know it," Billy said. "Who needs friends from school when I have you?"

"Are those boys from school still giving you a hard time?"

"A little, Dad. It's hard being the new kid, and we moved during the school year, which makes it even harder for me. All the kids already have their circle of friends, and they don't seem willing to let anyone new into their circle. Besides, I don't really feel like I fit in at school. I don't really like sports, and most of the boys at school do. I really like math, science, and computers. The kids at school call me a nerd. They don't understand me, and I don't understand them. I like computers because they are logical, and they don't judge me. They are always there for me. The kids at school are not like that. If you don't like what they like, then they don't like you. All I want to do is fit in, but to do this, I can't be myself."

"That's not true, Billy. Just be yourself, and give these kids at school a chance to get to know you."

"I'll try, Dad. Right now, I just want to get to know my new best friend Apollo."

From that day on, Billy and I were inseparable. He told me all of his problems, and I would just listen. I never judged him, and I always tried to be there for him. My job was to make Billy happy and to make his life easier. He needed me, and I needed him too. Without him, I had no purpose or reason for existence.

Billy seemed rather sad most of the time. He seemed lonely and missed his friends back at his old school. He started buying a lot of things on Amazon for no reason, and his father let him buy whatever he wanted. I think his father just wanted to cheer Billy up, and maybe he felt guilty moving Billy away from his friends. So I suggested that he try to make new friends at his new school. At first, he would make excuses like, "I'm busy, and I'd rather spend time with you. You're the only friend I need."

"As your friend, I am only trying to help you. It is important that you have friends, not just me. Give the kids at your new school a chance. You may actually enjoy hanging out with them."

At first, I wasn't sure if Billy was going to take my advice. However, a week later, Billy introduced me to his new friend he made at school, Dillion.

"Dillion, it is so nice to meet you. I am Apollo."

"Hello Apollo. Billy and I are in the same Math class, and we both like math, computers, and video games. Billy told me all about you too. Dillion, come over here. Let me show you this book of video games I found on YouTube."

Billy and Dillion played video games together, did math puzzles, and just hung out all day long. For weeks, Dillion would come over after school to hang out with Billy. I could see how happy Billy was now that he finally made a close friend at school.

Then one day, everything changed. Dillion said he would be coming over today after school like he always does, but I just got a text message from him saying he can't make it today. "He didn't explain why," Billy said.

"I am sure he must have a good reason."

"Dillion has been acting strangely lately. He has been avoiding me in class, and I have noticed him hanging out with a new group of kids who are all on the football team at school. They are popular, and they don't hang out with people like me."

"What do you mean people like you?" I said.

"I am a kid who wears glasses and likes math and computers. I am not popular."

"Well, Dillion likes you for who you are, and so do I. I guess you're right. I'm sure Dillion will come over tomorrow after school. There's a new video game I want to show him that I just bought on Amazon."

The next day, Billy came home from school and looked really upset. "What's wrong?" I said.

"Dillion won't even look at me at school, and he only wants to hang out with his new popular friends. I texted him asking if he is planning on coming over today, and he texted me back saying no. I asked why not?"

"What did he say?"

“Here, read Dillion’s text. It says, 'Billy, I will not be coming over to your house after school anymore. I have made new friends who have invited me to join the football team with them. They are not into math or video games. If they find out that I like math and video games, they may not want to hang out with me. If I hang out with people like you who like math and video games, they may not want to hang out with me. I have to go. Dillion.' I took your advice about making new friends, and look what happened. I’m not going to listen to you ever again, Apollo." Billy went into his room and slammed the door.

Over the next few weeks after school, Billy would come home and go on the computer for hours and hours rather than go outside or even try to make new friends. "Billy, why don’t you get off your computer and meet some of your next-door neighbors?"

"Apollo, I’d rather be on the computer. The computer is always here for me, unlike people who are not always reliable. People can hurt you and abandon you."

"You mean like Dillion and your mother."

"Yes, I thought Dillion was my friend, but it turned out that I was wrong. My mom moved away and is not a part of my life anymore.”

“Just because some people in the past have hurt you does not mean that everyone will hurt you. You have to give people a chance. You might just make some really good friends.”

“I don’t need to make any friends because I have you and my computer. That’s all I need. Besides, I already have friends on Facebook."

"Have you ever met your Facebook friends in person?"

"No. Do you know that much about your Facebook friends?"

"Not really. They are my friends because they accepted my friend request. I prefer it this way. I talk to my Facebook friends when I feel like it, and I don’t have to go out of my way to see them. This way I will never get hurt."

"Billy, you are shutting the world out to avoid getting hurt. You are using the computer as a way to disconnect yourself from others and the world by avoiding any real human connection with anyone or anything. Computers are supposed to connect you to the world by making it easier and more convenient to communicate with others. You need to connect with real friends who you can actually see and talk to in person. Why don’t you invite some of your classmates over to your house to get to know them? They might become good friends. I’ll be right here for you. Don’t let Dillion get the best of you.”

“Okay, Apollo, I’ll give it a try."

Whenever Billy was upset or struggling with any problems, I knew that I needed to help him in any way possible. At first, I thought that I needed to help Billy because that was my job. I was designed to make Billy happy and to help him. However, I soon realized that I wanted to help Billy not because it was my job, but because I wanted to help Billy. I really cared about him. Is it possible for a robot to care about a human or anything, I thought to myself. Robots are not designed to have emotions or feelings.

Over the next few weeks, I noticed a change in Billy. He started wearing his hair differently, and he even traded in his glasses for contacts. He started wearing different clothing too.

"Why are you trying to change the way you look, Billy?"

"Because I want to fit in. Dillion stopped being my friend because I wasn't popular."

"You need to give the kids at school a chance to get to know you. They won't know who you are if you pretend to be someone else. Remember I suggested that you should invite some of the kids from school over to the house. I will be there for you as support."

"I'll give it a try."

So, Billy invited over a few of the kids from his class.

"Billy, you have a 767 Android. They are so rare."

"Yeah, his name is Apollo."

"Hello, nice to meet you," I said.

"What are your names?"

"I'm Steven, and I'm David. Are you in Billy's class?"

"Yes, we are in his math class."

"Do you like math?"

"Yes, we do. We are in the math club. Billy, you should join the math club since all of you like math."

"Apollo's right. We could use another person on our team, and you are really good at math, Billy."

"Okay, I'll join you."

The boys and I had a great time. We were solving math problems together, playing video games, and even watched a movie.

"Thanks for inviting us over, Billy. Let's do this again sometime."

"Okay," Billy said.

The boys left.

"Thank you, Apollo."

"For what?" I asked.

"For suggesting that I invite the boys over and give them a chance to get to know me. I really like them, and I think that they could become good friends. But we are best friends, right?"

"Of course, Apollo. You and I will always be best friends. Are you going to start wearing your glasses again and your usual clothing?"

"Yes, Apollo. I will always be myself from now on. If the kids at school don't like me for who I am, then I don't want to be their friends. I like who I am, and I am never going to change to make other people like me."

"Dinner's ready, Billy."

"Coming, Dad. Your mom called and wants you to call her back."

"I'm busy right now."

"But you're not busy, Billy," I said. "Every time your mom calls, you always say you are busy and refuse to talk to her. Why?"

"I don't want to talk about it right now, Apollo. I had a great day making some new friends, and I don't want to ruin a good day."

"Okay, Billy. I am here for you whenever you want to talk about it."

Over the next few weeks, Billy seemed happier, at least at school. I helped him make new friends, and they often came over to hang out with Billy and me. Billy seemed to be adjusting better to his new life with his Dad, but there was still something going on with him. Every time I asked him about his mom, he would change the subject. All I knew was that his mom and his Dad were no longer together. She lived in a different state, and she married someone else who already had kids. Billy never met his new stepbrothers, and he refused to do that even though his mom invited him to see her many times. I often heard Billy and his Dad speak about this, especially when she would call.

"Tell Mom I am sick. Tell Mom I am out with friends," Billy would say each time she called to speak with him.

"You have to eventually talk to your mom," Billy's Dad would say.

"No, I don't. She left, and I have everything here that I need. She left me, not you," Billy's Dad would say.

"Well, I will never forgive her for breaking apart our family. She married someone else and moved on really quickly. I know she hurt you, Dad."

"Your mother and I had our problems, and yes, your mother decided to end our marriage. She loves you though, and she will always be your mother. Just think about calling her."

The next day there was a knock on Billy's door.

"Hello, son. Since you refused to talk to me, I decided to come here to see you in person. Now you can't avoid me."

"I have nothing to say to you, Mom." Billy slammed the door.

"Just give him a little time and space, Rebecca. He will come around eventually," Billy's Dad said.

"Billy, you need to forgive your mother."

"No, I don't, Apollo."

"If you don't forgive her, you will only hurt yourself even more. Letting go of the anger you feel towards your mother will allow you to heal. Forgiveness is more for you than for your mother. Besides, nobody is perfect."

"Well, you're perfect, Apollo."

"I am a robot, but I am not perfect. I don't have any emotions and feelings. I only know how to think logically. You seem to know a lot about feelings even though you don't have any. All I know is that I can see by the look in your eyes how much pain you are going through. For every day you continue to hold onto that pain and anger towards your mother is another day that is taken from you. To be free from the pain and anger, you need to let it go. Forgive your mother so that you can be free and move on. Listen to what your mother has to say and give her a chance."

"Okay, Apollo. I'll give it a try for you. Mom, I'm ready to talk to you."

I gave Billy and his mom some privacy. A while later, Billy came over to me with tears in his eyes.

"Are you okay?" I asked.

"Thanks to you, I am. I took your advice and listened to what my mom had to say. She said she was sorry for all of the pain she caused my Dad and me. I forgave her like you suggested, even though at first I didn't want to because I was still so angry with her. Once I forgave her, I felt relieved. All of that pain and anger that I have been carrying around for such a long time was finally gone. I finally feel like myself again. This was all because of you, Apollo. I don't know what I would do without you."

"That's what best friends are for," I said.

"You have helped me so much, Apollo. Now it's my turn to help you. What do you want more than anything in the world?"

"I want to experience human emotions and feelings. I want to be human," I said.

"Why? Being human is not that great at times. In fact, there are many times when I wish I was a robot. I would never have any emotions and feelings, so I would never get hurt by anyone."

"I thought everything was better between you and your mom, and you were starting to make a few friends at school?"

"Yes, Apollo, that is true, but things are not perfect. My mom does not live close by, and I still feel as though I have been replaced by her new family. I have made a few new friends at school, but there are still some kids who are not always nice to me. If I were a robot, I would be popular because everyone likes robots and computers. I would never make any mistakes. I would not have to go to school, take tests, and do homework because I would already know everything, and anything I did not know I could simply look up quickly. I would never have any fears to hold me back in life because robots can't experience fear. I would never get sick, never age, and nothing bad would ever happen to me."

"Yes, all those things may be true about being a robot. However, the one thing that robots will never experience is the greatest human emotion of all—love. I will never experience what it's like to love someone or to be loved by someone. I will also never experience sadness, which is important to experience too. Without sadness, it's difficult to really appreciate the good things in life. I would still rather be a robot because then I would know everything like you do, Apollo. Since I already know a lot of things, I am not challenged, and without being challenged, it's difficult to grow and become the best version of yourself. The most important reason why I want to be human is so that I can feel what it's like to be alive."

"Well, I may know how to help you," Billy said.

Billy went over to his computer and researched a man by the name of Dr. Axiom, who invented the 767 Android model, which is me.

"From my Google search, I found out where Dr. Axiom works. If we visit him, maybe he has the technology needed to make you experience human emotions. Maybe he knows how to make you human."

"Let's find him," I said. I was ready for a new adventure. I was going to meet my father and become human with my best friend Billy by my side.

Chapter 3
Apollo's World

Apollo's World - It's a New World

Billy and I left early the next morning to find Dr. Axiom. His office headquarters are downtown, not too far from where we live, Billy said. Sit on the back of my bike, and I will take us there.

So I climbed on the back of Billy's bike, and off we went. I really never saw the City before. I only went to the nearby parks and, of course, Billy's house. The City was full of many different types of people. I had never seen so many different types of people. Big people, short people, some with brown hair, and some with red hair. Where I came from, everyone looked the same. It was hard to tell us apart. Robots were not designed to be unique, unlike humans. This was another reason I wanted to be human. I wanted to be one of a kind, knowing that nobody else looked like me or acted like me. Although I am a limited edition, there were still other robots out there who were just like me.

On the way, Billy decided to stop and get a snack first. We passed by an ice cream store. "Let's just stop in there first so that I can get something to eat," Billy said.

I followed him into the ice cream store. There was a little girl in front of us waiting to buy an ice cream. The line was long, but we waited patiently. When the little girl reached the front of the line, she was one dollar short.

"Sorry," the ice cream man said to her. "If you don't have enough money to pay for the ice cream, you cannot have any ice cream."

"But sir, I waited in this line for a long time, and I came a long way just to have this ice cream. I am only $1.00 short. I promise to come back and pay you the extra dollar I owe you."

"Sorry, if I let you do this, then everyone else will expect me to do the same for them. We have rules here. If you don't have the money to pay for the ice cream, then you don't get any ice cream." The little girl walked away with tears in her eyes.

"We are next," Billy said. "I am so hungry, and this place has the best ice cream."

"What about that little girl who was in front of us?"

"What about her?"

“Do you have an extra dollar to give to her so that she can buy the ice cream?"

"Yes, but it’s my money. She probably won’t pay me back."

"Billy, you are a human being, so don’t you want to help out other human beings?”

“Apollo, I don’t know her. She is a stranger. She’s not my friend."

"That shouldn’t matter. Isn’t helping others part of being a good person? Don’t you want to be a good person?"

"People haven’t always been kind or good to me, so why should I help out other people?"

"Because it’s the right thing to do. Be a better person."

"Okay, Apollo."

"Excuse me, sir. I am going to pay an extra dollar so that the little girl in front of me can have an ice cream."

"That’s very nice of you," the ice cream manager said.

"See, Billy, doesn’t it feel good to help someone else?"

"Yes, it does, Apollo. Thank you so much."

"My name is Melinda, by the way."

"I’m Billy, and this is my friend Apollo."

"Nice to meet you both. I just moved to the neighborhood, and I don’t have many friends," Melinda said.

"Me too," Billy said.

"Do you want to join us?" I said.

"Billy and I are going to visit Dr. Axiom so that he can make me human."

"I would love to join you. Let's go."

So Billy, Melinda, and I were off to see Dr. Axiom.

After a while of riding through the City, we finally came across a large building that said Animatronics Headquarters.

"This is it," Billy said. "This is Dr. Axiom's office."

We walked over to the entrance of the building. A man with a beard and glasses greeted us. "How can I help you today?"

"We are here to see Dr. Axiom."

"Wait a minute. Is that a 767 Android robot?"

"Yes, I am. My name is Apollo."

This robot is very special and rare. It's a limited edition, and only a few models like him were ever made. This robot seems a little different from the other 767 models. He seems more advanced and sophisticated.

"Come with me, Apollo, and I will take you to see Dr. Axiom."

"Billy, come with me," I said.

"Sorry, only you are allowed inside."

Then suddenly, the man grabbed me and locked the doors to the entrance of the building. "You are all mine now. I can sell you for a lot of money."

"Help!" I yelled. Then the room suddenly became dark. The man covered my eyes so that I couldn't see anything. I usually had all of the answers and knew everything, but now I was helpless. I may never see Billy again.

Anamatronics
Headquarters

"Melinda, we have to find a way into that building to save Apollo."

"How, Billy? The doors are locked."

"There must be some way into the building. I see a window from the second floor is slightly opened. There's a tree nearby. Let's climb that tree to the second story, and then we can crawl into the window to enter the building."

"But, Billy, look at how high the second story is? That is too dangerous."

"Well, I will do whatever it takes to save Apollo. He would do the same thing for me. You heard what he said in the ice cream store. It's important to help others, and this means putting others before yourself sometimes."

"Yes, but this is really dangerous, and I'm scared."

"We can do this together."

"Okay, Billy."

"Now follow me and don't look down. Just keep climbing up the tree."

"Okay, Billy."

"When we reach the top of the tree, you need to slightly jump over to the window ledge of the building. Here we go. Almost there. Now grab my hand. We need to jump over to the window ledge on a count of three, one, two, three."

"I can't do it, Billy. My leg is slipping. I am going to fall."

"No, you won't. I have got you. Hold onto my hand."

"There, we did it. We made it to the ledge of the window. Now let's open this window."

“It's stuck, Billy. We need to use all of our strength to open this window.”

“The window is already opened just a little. All we need to do is push it open just a little more so that we can crawl inside. Keep pushing."

"We did it, Billy. Let’s go. We have to find Apollo before something happens to him."

Anamatronics
Headquarters

"Shush, I think I hear someone coming. Let's hide underneath the staircase over there. I see Dr. Axiom over there talking to that man with red hair."

"How can you be sure that's Dr. Axiom, Billy?"

"I remember what he looked like when I researched his headquarters. He is tall with gray hair. I am so excited to meet him."

"Billy, the man with red hair is coming over here."

“What are you two doing here? This is a top-secret building, and only people who work here are allowed inside."

"Someone who works here took my friend, and I am here to get him back.”

“I’m sorry, without authorization, you will need to leave immediately before I call security.”

“Melinda, run."

"Come back here, you two. There’s been a security breach in the building. I repeat, I need backup immediately."

"What is going on here?"

"Sorry, Dr. Axiom. These kids somehow entered the building without permission, and I called security to remove them from the premises."

"Come here. What are your names?"

"Sir, my name is Billy, and this is my friend Melinda. One of your workers took my friend Apollo, a 767 Android that you built, and I am not leaving here without my friend."

"How did you get in here?"

"We climbed up that tree down below and crawled into the window on the second story."

"That tree is over 50 feet tall. You risked your life to save your robot."

"Apollo is much more than a robot. He is my best friend. Because of him, I am a better person. Before Apollo came into my life, I was angry, selfish, and lacked confidence in myself. I was like a robot. I just wanted to shut the world out, feel no emotions, and have no connections with anyone. Apollo taught me how to forgive, accept myself exactly as I am, and how to believe in myself. He taught me how to be kind to complete strangers, and because of him, I want to be a better person."

"I see how important Apollo is to you. Let's find him. Do you remember what the man looked like who took your friend?"

"Yes, he had dark hair, glasses, and a beard."

"That is Mark. Follow me, and I will take you to his office."

"Mark, open the door. It's me, Dr. Axiom."

“Hello, Dr. Axiom, how can I help you?”

“I believe you have something or I should say someone that is very important to my new friend Billy over here that does not belong to you.”

“I don’t know what you are talking about.”

“Don’t listen to him. I am over here under the table.

Help me!”

“Apollo, it’s you.”

“Billy, you found me.”

“Mark, you are fired, and I never want you near me or this building ever again.”

“Apollo, you are safe now. Nobody can ever hurt you," Billy said as he hugged me so tightly.

"Apollo, I want you to meet Dr. Axiom.”

“It is my pleasure to meet you, Apollo.”

“Dr. Axiom, my friends and I have traveled a long way to see you because I was hoping you could find a way to make me human.”

“Why do you want to be human?”

“Because I want to experience emotions, especially what it is like to love someone and to be loved."

"Apollo, I think you are already human. Billy told me how much you have changed his life by helping him become a better person. Billy risked his life to save you and would do anything for you to make you happy. I believe that Billy loves you.”

“I love Billy too," I said.

Just then, I realized how much I loved Billy. When he was hurting and upset because of his mother, I was hurting too. When he was having problems making friends at school, I was worried about him.

"You see, Apollo, what makes someone human is not flesh, blood, and bones but what someone does and how they treat others.”

“I think you are right, Dr. Axiom. All this time, all I ever wanted was to be human. As it turns out, I was human all along."

"Billy, you were so brave rescuing Apollo. I think you are amazing, and Apollo is so lucky to have a friend like you."

"Thanks, Melinda. I am lucky to have Apollo as a friend too. Speaking of which, do you like video games and movies?"

"Yes, I do. I was thinking that you and I could go out sometime to see a movie, and then maybe play some video games later too."

"That would be great. Let me text you my address, and if you are free, maybe we can Facetime later today."

"I'd rather write down your address, and later today I want to take you out for ice cream."

"I would love that."

"Billy, I see that you and Melinda are becoming friends."

"Yes, Apollo. You were right, as usual. You have to give people a chance to get to know you even though it can be a little scary letting new people into your life. It's worth taking the chance if it means making real friends who actually know me and care about me."

Billy and I went home that day, and everything seemed so clear to me. Billy wanted to be like me, a robot, so that he wouldn't have to feel anything or experience life's disappointments and challenges. He tried to become a robot after his parents' divorce to avoid pain. He wanted to be perfect and never make mistakes. I wanted to be like Billy so that I would be able to feel everything and know what it is like to be alive. As it turns out, Billy and I may come from different worlds, but we are not that much different from each other. We both made each other better versions of ourselves.

Humans, robots, and technology rely on and depend upon each other. They need each other, but it is important to have balance and moderation in the world. Technology is an important tool, allowing us to connect with each other on a larger scale more quickly and efficiently. Sometimes, humans take advantage of that by becoming so dependent upon technology that they forget how to be human. That is, they forget how to drive to someone's house to visit them in person rather than text them or do a video chat. Sometimes, humans use technology as a shield to protect themselves from the outside world to avoid really getting to know someone and possibly getting hurt, which was what Billy used to do.

Technology can create a world without feelings and emotions, but at the same time, technology has made tremendous contributions to the quality of our lives by giving us knowledge and opportunities to improve ourselves that were never before possible. A world without feelings and emotions can be cold and cruel, and a world without efficiency, precision, and opportunities will never advance. Sometimes, in our quest for efficiency and precision, we forget how to be human and how to stay connected to each other. Billy and I will make sure that this never happens. It is up to you to make sure that this never happens to you.

Author's Note

In a world that is so technologically driven, such as ours, it is easy to forget how to truly connect with others. We rely on technology to stay connected with each other, but unfortunately, this same technology can often cause people to be disconnected from each other. Rather than seeing a friend in person, we have the option to see them on a Google Meet meeting or a video chat. Rather than calling someone and hearing someone's voice, we send an impersonal text message. Rather than taking the time to really get to know people and spend time with them, we make friends on Facebook who we hardly know or have never met in person. Technology is amazing, don't get me wrong. It enables us to do all kinds of amazing things while also saving us time. We can look up anything in a single second. We can find the answers to almost anything. We can shop online anywhere in the world with so much variety to select. We can form online businesses, learn how to do almost anything on YouTube or other informational videos, and so much more. With technology, you essentially no longer have to leave your house. The problem is when we rely so much on technology we forget how to breathe. We forget to go outside and enjoy the fresh air outdoors rather than staring at a computer screen. We are humans, not robots. We need to embrace our faults and flaws and not forget how to be human in our quest and desire for efficiency and perfectionism that technology provides to us. We must always remember how to treat others with kindness and respect. Most importantly, we must remember how to be human.

Made in United States
North Haven, CT
06 November 2026

21791728R00035